Things in The Ocean That Can Kill You

An introduction to the serious science of deadly ocean organisms

Bruce J. Stephen

Copyright © 2019 Bruce J. Stephen.
All rights reserved.

ISBN: 9781690027584

Contents

Acknowledgements	i
Section One — Introduction	1
Toxins	5
Teeth	7
Stinging Cells	11
Section Two — Don't get stung	
Box Jelly	17
Cone Snails	21
Flower Urchin	27
Man O War	29
Stonefishes	31
Section Three — Don't get bitten	
Blue-ringed Octopus	37
Sea Snakes	39
Sharks	43
Bull Shark	45
White Shark	49
Tiger Shark	53
Saltwater Crocodile	55
Section Four — Don't Eat	
Ciguatera	61
Pufferfishes	65
Red Tide Algae	67
Honorable Mentions	71
Sources and Further Reading	73

Acknowledgements

The concept for this book comes from lectures given at the Acadia Institute of Oceanography in Seal Harbor, Maine. Gratitude to them, instructors, students, and all.

INTRODUCTION

There are many dangers in the ocean, this primer focuses on the organisms that live there and how they might be dangerous. But first a caveat, this guide does not include all the dangers of the Oceans, it covers many of the most dangerous organisms. My goal is to enhance understanding and drive curiosity. Second, dangerous does not equal bad. Stop killing sharks: Without these majestic animals' ocean systems collapse and us with them. Third, I've included many images, but they are not too detailed. Most are line drawings, feel free to color them.

In the world of science, it is becoming practice to capitalized the accepted common name, White Shark rather than white shark, though some do not agree with this, I like this convention and use it here.

Finally, if you do something stupid, it's all on you. This book is not a substitute for medical treatment, or for what you should eat in the ocean. Be responsible for yourself.

A side note on names:
Binomial names (scientific names, AKA, latin names), are all references to the two-word name given to each species: The first word is the genus the second the specific, or species, epithet. Humans are *Homo sapiens*, (this means man (*Homo*), wise (*sapiens*), or wise man). The Green Sea Urchin is *Strongylocentrotus droebachiensis*, (this is the longest scientific name I know). This name means round (strongylo), spine (centro); the species name refers to the town of Drøbak in Norway, where, presumably, these urchins were found in abundance.

Things in the ocean that can kill you

Let's get to the killing. The two main ways to get killed from things living in the ocean are teeth and toxins: If you get bitten by something with big teeth you can, of course, bleed to death. You can also get so hurt by the bite (or are held down by the creature) that you drown. That does not sound like a fun way to go.

Toxins can get into your body from a bite, a sting, or you can eat something containing them. When you eat something toxic we call that a poison, when it is injected into you we call that a venom. Both can kill you. The actual toxic substance, whether eaten or injected, might be the same chemical.

Toxins are a growing problem for us as we continue to add pollutants to the Oceans. Many problems stem from an overabundance of toxic algae. The explosive growth of these algae are primarily due to the nitrogen and phosphorus pollution that runs from farm land into the water.

Algae, toxic and otherwise, are eaten by things that we, in turn, eat. Thus the pollutants that we dump into the oceans come back to harm us. Toxic algae are found all over the world but many other things that kill directly, by stinging or biting, are concentrated in the indo-Pacific. This region contains northern Australia, the Papua New Guinea, Malaysia, The Philippines, into Thailand, and Vietnam (see map).

Things in the ocean that can kill you

This map illustrates the general area known as the Indo-Pacific. The boundaries of the Indo-Pacific are not exact but its midpoint is about the center point of this map and includes northern Australia and extends north. It encompasses many islands and nations, only some are listed.

Things in the ocean that can kill you

Things in the ocean that can kill you

Toxins
The primary way things in the ocean kill you

The two ways toxins get into your body is 1. you get it injected from a bite or sting or 2. you eat the toxic animal. The jellies or jellyfish, are not the kind you put on a peanut butter and jelly sandwich, so they can kill you by injecting you with venom from their stinging cells. Note, peeing on a jelly sting does not help!

Most toxin work by disrupting some body function. Lots of disruption = sick, more disruption = death. Other toxins cause cells to break apart; for example, a hemotoxin destroys red blood cells (turns out they also destroy other cells and disrupt things like blood clotting too, but you get the idea), and a Myotoxin causes damage to muscle tissue.

Many toxins affect our nervous system cells, the neurons (these are neurotoxins). These cells are responsible for sending and receiving signals all over the body. This includes signals that tell a muscle to move, a signal to the brain that the stove is hot, and all sorts of internal signals, like those from the stomach to the brain, and in turn to glands, that release digestive enzymes in preparation for digesting food. If the signal is stopped, or even slowed down, it will disrupt the process: food isn't digested and signals do not get to your where they should. If signals are disrupted for too long, or completely stopped, it is deadly. An easy example that illustrates the importance are the signals that tell the heart to contract. If the heart stops contracting, blood stops flowing and cells do not get enough oxygen. Guess what happens to the cells then? And you shortly after?

Toxins come in all levels of potency, tetrodotoxin (a common toxin found in many organisms) is extremely toxic. The Material Safety Data Sheet for tetrodotoxin lists a dose lethal to 50 percent of mice as 334 µg per kg. A microgram, µg, is one millionth of a gram. That conversion doesn't help me to understand how small that is I just understand that a very small amount can kill you. This may help, the amount that would kill an average person is about 1/12 the weight of a penny. Cut a penny in half, then in half again, and again, and again, and again, and again and **this** amount will kill you.

FYI: You don't really need to cut a penny in half. ☺

There are so many toxins out there, some of they include ciguatoxin, saxitoxin, and brevetoxin. Even chemicals not considered toxins can be dangerous. For example, some cone snails inject their prey with insulin, a normal body hormone. The insulin level is so high that it causes the prey to be temporary paralyzed—and during this time the snail eats it.

Things in the ocean that can kill you

Teeth

These can kill you, but usually they have to be attached to an animal. When we think of animals with teeth that can kill you, that live in the ocean, sharks are first on the list.

The epitome of a shark tooth is large, sharp, and triangular; the type seen in the White Shark *Carcharodon carcharias*.

Generic shark tooth.

The reality is that there is no epitome of a shark tooth. Shark teeth vary by species, and position in the mouth (called heterodonty). The white shark, with its large, sharp, triangular teeth doesn't have the same type of teeth throughout its mouth. Those stereotypical triangles are found on the top jaw, toward the front, (subsequently they are the ones you'll see just before getting bitten) while the bottom is fixed with more pointed teeth. Sharks have lots of teeth, the nasty predator variety particularly. The White Shark has 48 teeth in the front row, and behind are 5 or more rows of teeth ready as replacements. That's at least 240 teeth at the ready!

All sharks are not nasty predators, and thus many sharks, and their close relatives, have very reduced teeth or teeth modified into crushing plates.

Things in the ocean that can kill you

Tooth from a Sand Tiger Shark. The Sand Tiger Shark has teeth with cusps that allow skewering of slippery prey.

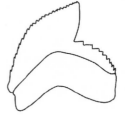

Tooth from a Tiger Shark. This is the best looking shark tooth: Serrations make the tooth.

The oddest shark teeth are those of the Cookie Cutter Shark. They are not odd because of their shape but because of their relative size. The cookie cutter shark has large teeth in its lower jaw, shaped much like those in the upper jaw of the White Shark. In its upper jaw the teeth are pointed and much smaller. The teeth of the lower jaw appear far too large for the shark. These little buggers (typical size of these sharks is under 30 cm) can take chunks of flesh out of prey—like a cookie cutter has been punched through the dough of tissue.

Cookie Cutter Sharks live in the deep ocean but they do travel around, moving into shallower water to hunt for prey. They take bites out of larger prey but I have never heard that they have ever attacked a person. Thus they don't really belong in this guide

Things in the ocean that can kill you

but they have such cool teeth.

The best fun fact about these sharks is they are known to have taken bites out of, wait for it, White Sharks.

○ ○ ○ ○ ○ ○ ○ ○ ○

The other major toothy predator in this guide are the salt water crocodiles. Their teeth are not serrated or as sharp as shark teeth. But their strong jaw muscles allow the croc to ram these teeth into their victim and once that happens getting away is dodgy.

A few teeth from a crocodile, showing the conical shape.

Other ocean denizens have teeth, of course, the sea snakes for example. But the danger in these slithery animals is not the teeth themselves, which are often so small they cannot penetrate a wetsuit, but the venom injected.

Things in the ocean that can kill you

A quick tale to so you do not misjudge anything with a mouth. Porcupine and balloon fishes do not have sharp nasty teeth, instead they have crushing plates used to eat hard shelled prey. Knowing this, a diver stuck his finger in the mouth of one. What the diver did not realize is that before the crushing plates is a set of sharp beak-like ridges. The diver lost part of his finger in that rather stupid encounter. FYI, the diver was not me, and I do not admit to knowing him.

Let's make a plan;

No matter what type of tooth, sharp, serrated, conical, even small, the plan, when you visit the ocean, is;

Do not get bit.

Cnidocytes- Stinging Cells

Within the tentacles of jellyfish (maybe we should call them jellies since they are not really fish) are specialized cells known as cnidocytes (the *C* is silent). These cells house a spiny tube, that may be called by several names, cnida, nematocyst, or cnidocyst.

The cell is called a cnidocyte, and the structures inside are cnidae (plural for cnida); some specific types of cnidae are called nematocysts. Cnidae that penetrate are called penetrant nematocysts, while others are spirocysts or ptychocysts (the *p* is silent, so its *tic o sist*). The nematocysts variety are the most familiar and the ones considered here. They are the pesky barbed tubes that are the reason people don't like jellyfish.

The cells that contain the stinging barb, are very tiny, microscopic. So if you want to see them bring out the microscope. The phylum that contains jellies, Cnidaria, is named for these cells (just like with the name of the cells the *C* is still silent).

The venom, delivered via stinging cells, immobilizes prey so the delicate jelly can safety start ingesting it. A thrashing fish could easily damage the jelly.

This cnidocyte is at rest, but primed for firing. On the top left side is the trigger (called a cnidocil). Inside, is the coiled tube that turns inside out when "firing". This cell is a trap ready to spring. Even the cnidocytes of a dead jelly can 'fire' and sting you. When the tentacle brushes up against prey, or your skin, movement of the trigger causes, among other things, the release of calcium ions into the cell. This, in turn, causes water to rush in due to osmosis.

Things in the ocean that can kill you

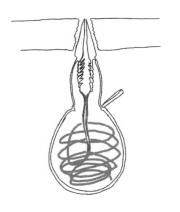

The pressure of the water rushing in causes the inverted tube to evert out the door (the flat space on top of the first cell, now pushed to the side). Along the inner membrane of the tube is the venom. As the tube everts the inner lining starts to become the outer layer. In this drawing, the charged cell has just begun releasing the nematocyst.

This penetrant nematocyst is now more emerged, everting. Barbs will stick into and on the prey. In this image barbs have penetrated the tissue (at top) allowing the tube to project into the prey. Keep in mind this is a microscopic view; the cell is about 30 microns across—3 cells lined up would span about the width of this line: |

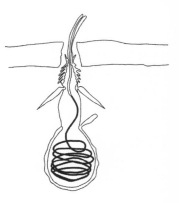

A fun with word roots aside: The nema root means thread (Greek origin), while cnida is Latin for nettle. In the word cnidocytes (cyte means cell). Science often has several terms for the same thing, one Greek in origin and one Latin. Consider the hormone that increases heart rate, called epinephrine or adrenaline, these two names refer to the same hormone. Named because it is released from the adrenal glands, which sit on the kidney (ad = near; or next to, renal = kidney. This is the Latin word root version. Whereas epinephrine means basically the same things (epi = on, nephros = kidney. From the Greek word roots).

Things in the ocean that can kill you

Other members of this phylum, corals, and sea anemones, also have stinging cells. Most of these have such tiny stinging cells that they will not hurt you if you touch them. The Cnidarians are a primitive group. Yet these relatively simple animals produce toxins that form holes in cells, or act as neurotoxins, vasodilators, or enzymes that destroy tissues. A huge array of different types of chemicals. Amazing considering the status of cnidarians as "simple" among the animals. Maybe they aren't so simple after all.

Another side note: The term jellyfish is most often used casually to describe the animals with stinging cells in the phylum Cnidaria. I prefer to just use the term jellies. Sometimes people use sea jellies. In addition, one group of jellies, the scyphozoans, are referred to as the true jellyfish. Nothing but confusion ☹

Side note, yes another one: From the animal-using-another-animal department: Some species of nudibranchs (sea slugs) can consume and use stinging cells for their own protection. These shell-less marine snails eat tentacles of jellies. Then they manage, somehow, to get the cnidocytes safety (that is without firing) up into the soft projections on their back (these are called cerata). The stinging cells are then used for protection, firing into potential predators of the sea slug. (drawing on the next page).

Things in the ocean that can kill you

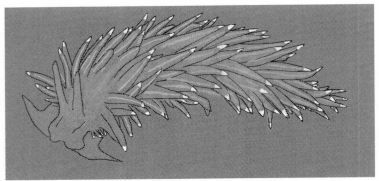

Nudibranch. Projections along the back (dorsal side) are cerata. The tips of these projection contain cnidosacs where nematocysts, that where originally consumed from another animal, may be housed.

The next time you're at the beach do not eat the jellies and try to sequester the nematocysts for yourself.

Things in the ocean that can kill you

Section Two

Do not get stung

Things in the ocean that can kill you

Box Jellies

Some jellies (are we still calling them jellyfish?) can be very small and almost invisible in the water. Others are easier to spot due to size and/or pigment. Many can give you a nasty sting and the deadliest are members of the box jellies. One species, *Chironex fleckeri*, known as the Sea Wasp grows larger than most, reaching up to 30 cm in diameter (about the width of a piece of paper). The Sea Wasp's tentacles, however, can be over 3 meters long (more than twice your height—unless you're really tall—are you really tall?). You could run into the stinging tentacles without even seeing the rest of the animal. These box jellies, as the name suggest, are cube shaped. Jellies in this group are known as the Cubozoans (cubed animals) and have tentacles emerging from each bottom corner of their cube. There can be more than 15 tentacles from each corner of the Sea Wasp.

Another deadly box jelly is the Irukandji jelly, *Carukia barnesi*. These are small jellies reaching a maximum size of 3 cm (about a tenth of the size of a Sea Wasp). They have a single tentacle (up to 6 cm long) in each corner of its 'box'. If you get stung you may get Irukandji syndrome, which is a whole host of symptoms including nausea, muscle pain, sweating, and high blood pressure. The symptoms do not begin immediately after the victim gets stung and may take 2 hours to develop.

The syndrome is named after the indigenous Australian people, from Queensland, the Yirrganydji. The species epithet (second word) in its binomial name *Carukia barnesi* refers to Jack

Things in the ocean that can kill you

Barnes, the discoverer of the jelly. There is a story that Jack let himself be stung by one of these jellies to determine if it was the cause of Irukandji syndrome. If true, that is dedication to your science.

Like the Man o' War, included in this guide, box Jellies are Cnidarians (Phylum Cnidaria). While the Man o' War is considered a colonial organism (cells that cooperate and live together) box jellies are considered individuals. That is, they have, for example, a digestive system, primitive though it is, that works to feed the rest of the body. The Man o' War does not have a digestive system and instead relies on specialize individual cells for this function.

One commonality between these different types of Cnidarians are the stinging cells. The stinging cells, cnidocytes, that cause the problems, contain a whip-like tube (like a garden hose)—see the cnidocyte section at the beginning of the book. Inside the hose is the toxin. As the tube comes out it turns inside out presenting the toxin to the prey (or your arm).

Until recently there was no antidote for box jelly stings but recently scientists from Australia have produced a possible treatment. It can stop the sting, and at least some other symptoms, if applied to the area within a few minutes.

Things in the ocean that can kill you

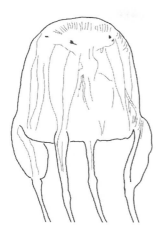

The bell (top) portion of an Irukandji box jelly. Its tentacles can be more than twice the length of the bell. It has a single tentacle extending from each bottom corner.

The Sea Wasp can have up to fifteen, 3-meter tentacles extending from each corner of its cube.

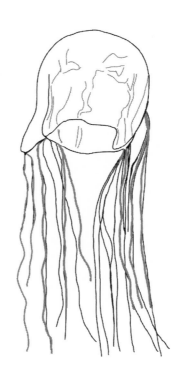

Things in the ocean that can kill you

Cone Snails

If you dip your hand into the warm shallow waters of the Indo Pacific Ocean you might chance upon a large multicolored shell of the Geography Cone Snail, *Conus geographus*. If you pick it up, this large shell, reaching to more than 15 cm (6 inches), is surprisingly heavy. It has a dense thick shell. The outermost layer of the shell is made of protein, called the periostracum, and this protects the shell. The periostracum also prevents the shell from being smooth, so when you run your hand along the surface of the shell you will feel the pattern ripple beneath your finger.

If, when you pick it up, its long snout (proboscis) extends it might impale you with a tiny dart. If that happens, well, you shouldn't have picked it up, because this is a cone snail, a type of predatory venomous snail. Among the cone snails, the venom of this particular species is the deadliest. Its dart is modified from the radula teeth (toothed tongue common in snails). A snail's radula is usually used to scrap algae from rocks or, in predatory species, drill into the shell of a clam or mussel. This radula, modified into a dart, is different. The dart is wrapped, like a rolled up a piece of paper, the hollow space inside is filled with toxic venom, and at the end is a hook. The strike of cone snails, it was discovered recently, is one of the quickest found in the whole animal kingdom. Using a mechanisms of a latched-doorway like opening along with muscle power, they propel the venom soaked radula hydraulically. Speeds of up to 400,000 meters per second (more than 800 thousand miles per hour) were measured.

Things in the ocean that can kill you

Cone snails are in the phylum Mollusca (which mean *soft-bodied*), class Gastropoda (which means *stomach foot*). There are over 80,000 species of gastropods, this includes all snails and slugs from the giant African land snails, to delicate and colorful sea slugs of the coral reefs. besides cone snails, other snail species use a toxic dart, like the auger shaped terebrid snails. They are related to cone snails but look much different. Some land snails use a dart during sexual foreplay—yes, it is called a 'love dart'. The dart contains a hormone that enhances the possibility that sperm fertilize the eggs. The development of this dart is different from that of the cone snail darts, not being modified from radula teeth, but instead are calcium carbonate arrows made within the reproductive system of the snail. Thus they appear to have different evolutionary origins. Speaking of origins, why is cone snails venom so toxic?

Most cone snails feed on worms, and the venom of those species isn't as nasty toxic as those that feed on fishes. If you are a snail (slow) preying on fishes (fast), you need to stop your prey quickly. It makes sense then, that the prey of cone snails determines how toxic their venom is. The Geography Cone Snail feeds on fishes. Still, you don't want to get harpooned by even the milder cone snails, the venom would disrupt the function of your nervous system. The toxins, called conotoxins, come in many forms and can interrupt the flow of nerve signals by blocking sodium, potassium, or calcium channels as well as getting in the way of receptors at the end of nerve cells.

Since many cone snail venoms interrupt nerve signals, they may offer a pharmacological tool as pain killers. One drug from conotoxins is now available, called Ziconotide, and others are being developed. These drugs may provide a less addictive pain medication then, for example opioids, which have caused so many addiction problems in recent years.

Darts of two different species of cone snail. The top image shows a full dart while the bottom image is a close up of the tip of a dart. At the microscopic level the dart is like a piece of paper rolled up into a thin tube, this tube is filled with venom, and, of course, it has a barbed point to pierce flesh.

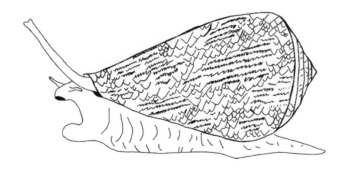

Geography Cone Snail, Conus geographus, is the cone snail with the most potent venom. Its shell is marked with lines, triangles, and patches of shades of brown. The extended siphon brings in water to pass over the gills and acts as a sense organ.

Things in the ocean that can kill you

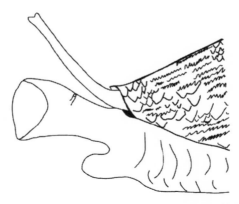

Another view of Conus geographus with its mouth extended. This snail can reach lengths of over 15 cm. It has eyes on small stalks, which are on the side of its mouth. The mouth can extend into this large funnel, which allows it to swallow prey whole.

Some cone snails use insulin, or a chemical that mimics insulin, to immobilize their prey. Insulin is a hormone that helps blood sugar, glucose, move into cells where it is used to make the energy molecule ATP. By flooding their victims with insulin the cone snails send them into low sugar shock (Hypoglycemia). This same shock occurs in a person who is diabetic if they took too much insulin. Taking too much insulin can be deadly and causes all sorts of symptoms. The symptom the cone snails wants to illicit in a fish is to prevent it from swimming away, enough insulin will cause a fish to be immobile. Ugh, this means the fish, not yet dead, watches itself being slowly enveloped by the snail.

There are many hundreds of species of cone snails and most live in the warm waters of the Indo-Pacific. But be wary, tourist havens like Hawaii house cone snails and some species live in even cooler waters.

Things in the ocean that can kill you

My advice, don't get stung by a cone snail.

Quick procedure for if your dive buddy is stung:

First, it probably won't kill them unless they stay in the water, so…

1. Get them out of the water.
2. Get them to a medical facility as quickly as possible. The biggest problem is if the toxin prevents nerve and muscles action that causes them to stop breathing.
3. Do not bother to collect the snail to determine the species. There is no anti-venom for any species. And thus no reason to risk being stung yourself, and no reason to take the time to find it—go to the hospital.

Things in the ocean that can kill you

Flower Urchin

The Flower Urchin, *Toxopneustes pileolus*, looks like it is made up of hundreds of tiny flowers. But be wary, the "flowers" are not flowers, instead they are pedicellariae, which are tiny jaws on the end of stalks that project out from the body of the urchin. These often extend beyond the spines. They are protective devices that say to any would-be predators, 'stay away'.

Some pedicellariae are small and help keep the urchin clean, removing debris. Other pedicellariae, like the small circular flower-looking ones, have a barbed tip, that, you guessed, can deliver a venom. Don't touch, step, on, or otherwise mess with these urchins.

If you do step on one of these and you can survive for long enough, the symptoms of envenomation, including pain, loss of muscle control, and paralysis, are short-lived. In about 20 minutes, after you accidently step on one of these urchins, most of the symptoms will be gone. The key is surviving until then, so make sure you are out of the water or else you'll drown.

The venom is protein based and heat can help to break it down. If you get stuck, try some hot water to ease your pain. Not too hot, that will burn you.

Other species of urchin are known to have toxic properties, like the Long-spined Black Sea Urchin, *Diadema antillarum*, found in the Caribbean. In these species the toxin is in their spines not in the pedicellariae and isn't nearly as potent. In fact, many members of the Echinodermata, the phylum that contains urchins, sea stars, and sea cucumbers, have toxins. Some are harmful to eat (poisonous) and some have toxins within their spines. The Flower Urchin appears to be the only one with venom associated with its pedicellariae. Let me walk that statement back, I predict that we will find many other species with venomous pedicellariae, we just haven't found them yet.

The Flower Urchin is found all over the indo-pacific and all the way through the Indian Ocean to the east coast of Africa.

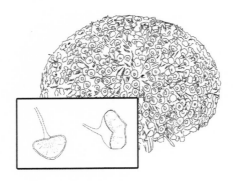

A Flower Urchin with an inset showing two of the dangerous flower type pedicellariae, one open and one semi-closed.

Things in the ocean that can kill you

Portuguese Man o' War

The Portuguese Man o' War is a Cnidarian. The Phylum Cnidarian, contains corals, sea anemones, and jellies (As you know I like to call them jellies rather than jellyfish). For their nasty, and potentially fatal, sting, the jelly's, like the Man O' War, are the best known.

The Portuguese Man O' war (scientific name, *Physalia physalis*) is actually considered a colony of cells. The cells work together in a type of symbiosis. Because of this it's classified in the order Siphonophora. The cells, although in some ways distinct, are so connected to the other cells that they cannot survive on their own. Some cells are feeding cells, some are stinging cells, that have nasty venom filled barbed tubes (see the front section on nematocysts) they use to sting and immobilize prey (or unwary swimmers).

The most prominent part of this jelly is the float. It is a beautiful iridescent blue. The float remains above the water and acts as a sail. Jelly's, including this one, often occur in blooms where many hundreds or thousands appear at the same time. These many floats must look like an attacking navy, like the other type of man-of-war, the old style sailing warship. This is, of course, where the name of this jelly comes from.

This jelly is responsible for tens of thousands of stings each year but, fortunately, deaths are rare.

Things in the ocean that can kill you

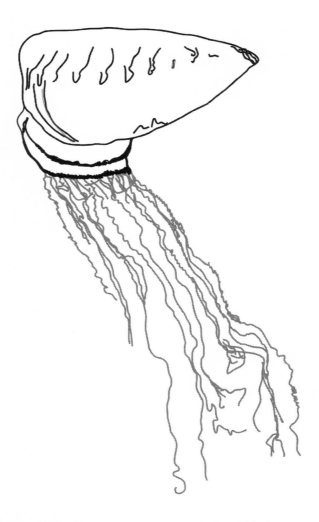

The Portuguese Man o' War is one of the most beautiful looking masses of gelatin you will find. It's also deadly.

Stonefishes

The first time I walked into the warm ocean water I was with a group of students. We were told to shuffle our feet as we walked into the deeper water. One reason to do this is that this 'scares' up the bottom denizens and so you won't step on them—many have spines and sometimes those also have toxins. Another reason for a whole group of students to shuffle their feet is because it looks kind of funny.

By far the biggest reason to shuffle your feet is the Reef Stonefish. The Reef Stonefish, *Synanceia verrucosa*, is probably the most toxic fish in the world. Its dorsal fin spines have twin venom sacs that contain a mix of toxins: the mix breaks down proteins (verrucotoxin), breaks apart blood cells (stonustoxin), and attacks heart muscle (cardioleputin). The one saving grace? They will not attack you. But it is found waiting in shallow waters exactly where unwary people venturing out from the beach step on it.

The Reef Stonefish is not aggressive; it lasy on the bottom waiting for prey to come to approach. Take note, however, they camouflage well and thus are hard to see. Due to their array of toxins if you get stung you'll experience pain, swelling (because your small blood vessels are leaking), and heart beat disruption (arrhythmias), that can lead to death. The amount of toxin needed to kill an average person is about 100 mg. That's only about two drops of toxin.

Things in the ocean that can kill you

A general stonefish spine showing twin bulbs, on either side of the spine, filled with venom. If you step on a stonefish your weight pushing down over the spine squeezes the bulbs and pushes the toxin into you.

Unlike many toxins mentioned here, stonefish toxins are unstable and break down quickly. You can speed the breakdown by heating it. One treatment for the stings is to apply heat. If you're so inclined, you can also eat these fishes as long as you cook them. However, once you see the number of spines these fish contain I'm not sure you want to actually eat them.

The Reef Stonefish as it would be sitting on the bottom waiting for prey. It has a large mouth that, when opened quickly, helps to suck prey in. More interesting, and more dangerous, are the dorsal fin spines that contain venom sacs.

Things in the ocean that can kill you

A side note: Among the fishes of the world, there are more than 25,000 species, there are many dangerous ones. With so many species we should expect this. There are actually more fish species than all the other vertebrate groups combined.

Another side note: Both fish and fishes are used for the plural of fish. The practice, in science, is to use fish, one fish, two fish, when referring to the same species, and to use fishes when referring to multiple species.

Back to the stonefishes (multiple species): The most dangerous species, the Reef Stonefish, is found in the Indo-pacific. Other species of stonefish, and their relatives, are found elsewhere, for example the Scorpionfish of the Caribbean. Though dangerous, Scorpionfish have a much less devastating toxin. You still do not want to step on one as you wade into the water so shuffle your feet.

Things in the ocean that can kill you

Section Four.

DO NOT GET BITTEN

Things in the ocean that can kill you

Blue-ringed Octopods

There is a story about an unfortunate vacationer that picked up a 'cute' octopus and put it on his shoulder to allow his wife to take a picture. It then "hardly" bit him… he was dead by the afternoon. I'm not sure if that story is true but it's a good cautionary tale.

These reef denizens are cute but deadly, the four species of blue-ringed octopus have nasty venom in their saliva. All are in the genus *Hapalochlaena* and they are 10-20 cm long. You could probably hold 4 of them in your hand—but don't. They are found, as with most things in this guide, in the Indo-Pacific region.

Octopods are in the phylum Mollusca, in the order Cepalopoda (which means 'head' 'foot'). Their close relatives are squid and cuttlefish. The have good vision due to their well advanced eyes, but apparently they cannot see color. Octopods are probably the smartest invertebrate. So far the octopods have used their brain power to hunt and catch prey and not take over the world, so we are safe. FYI; some use octopuses (and also octopodes) for the plural of octopus. I'm okay with that, but don't use octopi.

As I mentioned, the venom is in their saliva, so if you get bitten you're in trouble. The bite is just a tiny prick and victims sometimes do not know they have been bitten. The venom is a neurotoxin, tetrodotoxin (usually shortened to TTX), which effect neurons that control muscles, paralyzing the muscle. So once you figure out you're dying, you can't move to get help.

Things in the ocean that can kill you

There is no antidote to the toxin but at least a few people that were bitten have survived because they were put on artificial respiration until the toxin dissipated and they regained their own muscle control.

The toxin that kills you is actually produced by bacteria (housed inside the gland that produces saliva in blue-ringed Octopods). This same toxin is in other animals too; worms, a sea star, a salamander, and several fish species, including the pufferfishes: The family of pufferfishes, Tetraodontidae, is where the toxin gets its name.

Tiny but deadly. The blue-ringed Octopus has, you guessed it, blue colored rings, their body is brown or brown/yellow. The rings are iridescent, which means they change color a bit depending on the viewing angle. Be wary, the blue rings become more visible when the animal feels threatened; a warning display.

Things in the ocean that can kill you

Sea Snakes

The Yellow-bellied sea snake, *Hydrophis platurus*, appears to be the most widely distributed reptile species in the world. It ranges from the horn of Africa east all around the globe to the Isthmus of Panama. They miss covering the globe entirely by 11 thousand kilometers (between Panama and the horn of Africa across the Atlantic Ocean). This makes its east-west range about 25,000 kilometers. There are some sketchy reports of individuals occurring in the Caribbean Sea but there is no evidence of a breeding population. In fact, none of the species of sea snakes, about 68 species, occur in the Atlantic Ocean.

Sea snakes, or as my friend Bridget calls them, "Snea Snakes", are related to cobras (and were once placed in the same family, Elapidae, but now they are placed in the family hydroelapidae). Similar to cobras, most species have toxic venom.

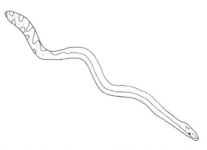

The Yellow-bellied Sea Snake, the most widely distributed sea snake, and probably reptile, in the world. It has a flattened tail that gives it swimming ability. Its head is at the bottom right.

Things in the ocean that can kill you

The most toxic sea snake is the Faint-banded Sea Snake, *Hydrophis belcheri*, also known as Belcher's sea snake. It can grow up to a meter in length. The species more dangerous to man is the Beaked Sea Snake, *Enhydrina schistose*, because it is encountered more often by fisherman—the group of people most bitten. Though the venom might be potent, deaths from sea snakes are unusual, probably due to their non-aggressive nature. There is an anti-venom for sea snake venom but even before anti-venom was available, the death rate from sea snake bites was low. In 2018 there was a death in Australia from a sea snake bite. This was the first death in the country for 80 years.

It was once thought the sea snakes could survive by drinking salty sea water, especially those species that spend time out in the open ocean. It turns out that these snakes actually drink from rainwater pools that form on the surface after rain. This bring to mind lines from that famous poem Rime of the Ancient Mariner by Samuel Coleridge:

> *Water, water everywhere, and all the boards did shrink*

> *Water, water everywhere, Nor any drop to drink.*

I wonder if sailors of old scooped water from pools at the surface when they were far out to sea.

Things in the ocean that can kill you

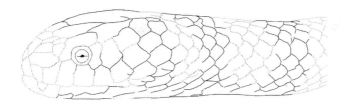

Faint-banded Sea Snake. This species is beautiful; it has black bands interspersed with yellow-white bands. Its eye, and the area directly around it, is an amazing blue.

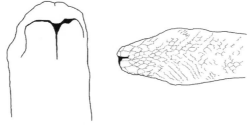

The Beaked Sea Snake has a cleft in its lower jaw, shown above in two different images from under the snake, one with scales outlined one without.

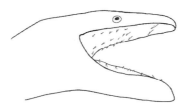

Sea snake head and jaws, showing small teeth. The snake's small teeth are hollow, and venom filled. This means that you could be envenomed by a dead snake. Be careful.

Things in the ocean that can kill you

Things in the ocean that can kill you

Sharks

Sharks are cartilaginous fishes, having skeletons made of cartilage and not bone. Though cartilage is more flexible than bone, your ears and nose are made of cartilage, (of two different types and different levels of flexibility), the overall body of a shark is pretty rigid. Sharks are very efficient swimmers, moving great distances without using up much energy, this is at least partially due to their body form, that cuts easily through the water, and this is helped by their scales (dermal denticles) that break up water turbulence. The fastest sharks are the mako sharks (*Isurus* sp.). The Shortfin Mako *Isurus oxyrinchus* currently holds the speed record.

Many sharks have a spiracle, a pair of holes that allow water to flow across the gills. This allows some species to sit on the bottom and still get oxygen from the water passing over their gills. Some sharks, the White Shark, for example, does not have spiracles. Perhaps, just perhaps, this shark needs to always be swimming to aerate its gills.

Sharks have a good sense of smell; our sense of smell is chemical detection in the air while theirs is chemical detection in the water. In addition to a good sense of smell sharks have an extra sensory ability, electroreception. That is, they can detect electric currents. They do this with organs called ampullae of Lorezini.

Things in the ocean that can kill you

Each ampulla are small sense organs usually found at the front of the shark. If you look at a close-up of a shark head you will see small openings (jelly-filled pores) that lead to the organs. This enables them to hone in on prey, but its only usable when the sharks get close. Sharks are not the only fishes with an electric sense, it is also found in ancient fishes like lungfish.

Many shark species have an extra eyelid-like membrane, the nictitating membrane, that cover their eyes when they go in for a bite. Other species, like the White Shark, roll their eyes back to protecting them.

Sharks have been evolving for over 450 million years giving rise to the current 400+ shark species. Shark fossils from the Devonian era (400 million years ago) have the same basic body form as todays shark and so these creatures are sometimes referred to as living fossils.

Sharks are beautiful, amazing animals. Let's stop killing them. By far the greatest conservation concern regarding sharks is due to their being over-fished. Most sharks are not dangerous and would have no place in here. Of course, this book being what it is, I've included the three species of shark that are considered the most dangerous.

Things in the ocean that can kill you

Bull Shark

The Bull Shark, *Carcharhinus leucas*, which is the same species as the Lake Nicaragua Shark, and the same species as the Zambezi shark (informally "zambi"), is one of three sharks in this guide. It, and not the White Shark, is responsible for most shark attacks on people. This is due to them hunting in shallow waters—exactly where beach goers are when they wade into the sea to cool off.

The Bull Shark is one of a number of fishes that can move from the ocean into freshwater systems, like Lake Nicaragua in, well, Nicaragua, where they may live for long periods in freshwater. They are also known to swim thousands of miles (the record is more than 2400 miles) up large rivers, including the Mississippi River (wouldn't it be cool to see a Bull Shark in the river in St. Louis Missouri, or in Iowa).

What features of the Bull Shark allow it to move, and live, in freshwater? For one, they pee a lot. When they move into freshwater they take on water by the process of osmosis, thus they need to get rid of this excess water. If you place most ocean fishes in freshwater they wouldn't be able to rid themselves of the excess water without getting rid of vital salts at the same time. The Bull Shark can do this, in part because its kidneys produce urine that is very low in concentration—mostly water, allowing them to rid themselves of the water. Under normal conditions, a shark swimming in sea water, sharks need to get rid of excess salts. Sharks do this via their rectal gland, a small worm like organ within their digestive systems. As you might expect, the rectal glands of sharks living in freshwater is reduced in size.

This view of a Bull Shark is the one you don't want to see if you are in the water. However, as with most shark attacks on humans, attacks by Bull Sharks seem to be mistakes by the shark. These sharks usually eat smaller fishes.

Things in the ocean that can kill you

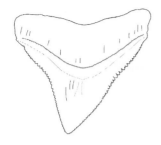

A Bull Shark tooth. On this tooth I included the bourlette, this is the area between the root and crown; the crown is the biting surface, serrated in this tooth, and the root is under the gum (top section in this picture). On the bourlette the enamel of the tooth is thinner because it is protected by a bit by the gum. Collectors of fossil teeth look for a good quality bourlette.

Bull Sharks grow to more than 3.5 meters (11.5 feet) and a big one weighs more than 200 kilograms (450 lbs, or two large people). They are muscular and powerful (which accounts for their name) and have a powerful bite, even among the sharks, though not as powerful as the Saltwater Crocodile bite.

Even with all the scariness regarding shark attacks, in the U.S. there are only about 15 attacks a year. In all of the planet there are only about 80 a year; shark attacks are rarely fatal. As people are fond of saying, you are more likely to be struck by lightning.

Out of nearly 500 shark species, three are responsible for most unprovoked attacks on humans: The White Shark, the Tiger Shark, and, you guessed it, the Bull Shark.

Things in the ocean that can kill you

White Shark

Few animals get the amount of press that the White Shark gets: Their agents must be really good. Their reputation is mostly due to the fact that they are big, have large teeth, and were the subject of one well known critically acclimated movie, Jaws. These features, but mostly the movie, all lead to its mystical and fearsome reputation.

The White Shark is more dangerous as it ages, young feed on fishes while adults transition to feeding on mammals, like seals. White Shark attacks on humans are usually thought to be mistakes: Does a paddling surfer look like a seal?

Shark have lots of teeth, the nasty predator variety, like the White has 48 teeth in the front row, with many rows behind filled with teeth ready to replace ones lost from the front row. At least in predatory shark species teeth grow and are replaced throughout their lives. How many in a lifetime? I can't find an analysis of this except on blogs (for example from the Sharkwatch SA Blog, Michelle Wcisel estimated a White Shark goes through over 29,000 teeth in a lifetime; the formula used was rate of tooth loss x average lifespan of shark (30 years) = teeth over a lifetime). She estimated the rate of tooth loss at 19 per week (988 a year), which seems very high to me. This value, about 30,000 teeth in a lifetime, was the same number I heard from a sea world educator the last time I was there, but I can't find actual data on the subject.

White Sharks are found throughout the world's oceans in places where the water doesn't get too cold. Australia, of course, is

Things in the ocean that can kill you

known for White Sharks, as is the tip of Africa. In the U.S. White Shark sightings are common on both coasts. On the U.S. east coast from mid-summer on, when the water is warmest, White Shark sightings are common along the beaches of Cape Cod.

Things in the ocean that can kill you

The White Shark, Carcharodon carcharias. These large predators, the largest confirmed is 7 meters (23 ft), has an even larger extinct relative, the Megalodon (Carcharocles megalodon). I know many refer to this shark as the Great White Shark, I prefer just White Shark.

Things in the ocean that can kill you

Things in the ocean that can kill you

Tiger Shark

I think we missed a great opportunity to call Tiger Sharks, Sea Tigers. Is it too late to change the name? Probably.

Tiger Sharks are, well let's be honest here, very cool looking sharks. They have stripes along their sides, thus the reason for the *tiger* moniker. These stripes fade and thus are less visible as they age. Too bad because a big ass shark with dark tiger stripes would be very cool.

Tiger Sharks are big, one of the larger sharks, reaching lengths to nearly 7 meters. And, no surprise here, they are one of the dangerous sharks. They are found around the whole planet in warmer seas, sometimes around coral reefs. Larger adult sharks are typically found offshore.

These fecund breeders (fecundity is the term used for reproductive potential), give live birth to lots of young pups. Most shark, but not all, give live birth. The mama sharks might have more than 50 pups at a time but still have a conservation status of *Near Threatened* because they are overfished, and many are caught as bycatch. They are also a target of the abhorrent fishery for shark fins.

The Tiger Shark, *Galeocerda cuvier,* is known to be a generalist feeder and even sometimes feeds on young tiger sharks. What the…. that just doesn't seem like a good idea for the species.

It is dangerous and is the culprit of some much publicized attacks on people in Hawaii years ago. It belongs to the family of sharks known as the requiem sharks, Carcharhinidae. A requiem is

Things in the ocean that can kill you

a Mass for the souls of the dead. Whoever named this group of sharks has a morbid sense of humor. These sharks have been seen ganging up on large prey, like whales, so the name is appropriate, but still morbid.

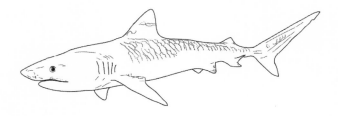

The Tiger Shark has light lines, which outline dark spots, on its back. These fade as they age. This is one cool looking sharks, it just is.

Things in the ocean that can kill you

Saltwater Crocodile

It has a prehistoric 12-meter-long relative, that might have feasted on dinosaurs. The relative is a bit larger than the currently living 6-meter length (20 feet) Saltwater Crocodile (*Crocodylus porous*), but that's still impressive. Seeing one of these ancient-looking beasts you can easily imagine their ancestors swimming alone the shoreline with dinosaurs walking close by.

The overall body form of the crocodiles has not changed much in the millions of years since they evolved—over 80 million years ago. The Saltwater Crocodile, that is the species *Crocodylus porous*, has been around for at least 4 million years. This is considered a living fossil.

The mouth and jaws of these crocodiles are huge, they have a wider snout than most crocodiles and because of this they look a bit like alligators. They also have the greatest bite force of any crocodile: If one of these beasts gets a hold of you, their teeth will be the least of your problems; their powerful muscles will crush your bones as they pull you under water and twist your around until you drown. This is not the best way to spend your afternoon.

Things in the ocean that can kill you

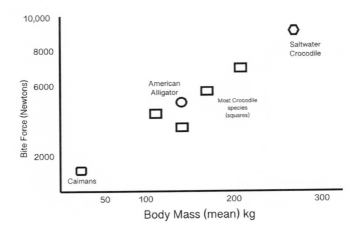

Here's a quick graph of bite forces in the crocodilians. The Saltwater Crocodile is all the way up on the right. (modified from the PLoS One paper by Erickson et al 2012.)

Some of the creatures in this book are unlikely to actually kill you, but Saltwater Crocodile's kill, about 3 people a year. They can swim long distances, and are often seen far offshore. They can also run very fast for short distances. Did I also mention that this is a big animal? This is not a beast to mess with.

At this time these crocodiles are not considered an extinction risk. However, this species range has been greatly reduced in the human era, which is called the Anthropocene. They are no longer found in some parts of Indonesia and are never found in China, though historically they once were. They suffer from hunting for skins or just because they are big and scary.

Things in the ocean that can kill you

They are deadly, yes, but we are smart enough to let them live their lives and to live ours.

Let's stop using the "its scary" reason to kill things.

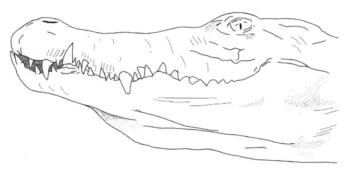

The Salt Water Crocodile has never been considered a beauty.

Things in the ocean that can kill you

Things in the ocean that can kill you

Section Five

Don't Eat

Things in the ocean that can kill you

Things in the ocean that can kill you

Ciguatera

Many fishes on the coral reef eat algae. These algae eating fish, in turn, are eaten by larger fish. Those fish are often eaten by even larger fish. This sounds like a Dr. Seuss book doesn't it? Wait for it. If the algae eating fish eat algae that is toxic, then we have a problem. One small fish eats lots and lots of algae. The algae eaters may not build up too much toxin but, you knew there was a but didn't you, but when many algae eaters are eaten by larger fish and in turn, many larger fish are eaten by even larger fish, the toxins accumulate inside the fish. The largest fish in the chain has the most toxin. This concept is called bioaccumulation. Thus be wary of large predatory fish as they are the most dangerous to eat: Barracuda, groupers, Red Snapper, and sea basses are a few examples.

One toxin is the ciguatera toxin, often called ciguatoxin, that is contained within some dinoflagellate species. Dinoflagellates are microscopic algae. Dinoflagellates with ciguatoxin are common in tropical seas, including in and around coral reefs. Other toxins might also be present in these microscopic algae. The toxin doesn't appear to hurt the fish, so you can't tell if a fish is infected.

Be wary when eating at the top of the food chain. Eating things that eat other things tends to be more dangerous.

    Can you cook the toxin out?
                No.

    Can you tell if a fish contains the toxin?
                No.

Things in the ocean that can kill you

These nasty but tiny algae are different than the nasty but tiny red-tide algae because they live on other things. So while red tide algae may signal an outbreak by coloring the water, the presence of large numbers of algae that cause ciguatera is hidden.

Ciguatera toxin disrupts signals in the neurons, muscle, and our cardiac muscle (heart muscle) cells. It produces gastrointestinal distress; nausea, vomiting, pain. This leads to worse symptoms including coma, slow heart rate, and (if you get a really large dose) death. Fortunately poisoning of this kind rarely leads to death. Unfortunately, some symptoms, like muscle weakness, can last for many weeks after the poisoning.

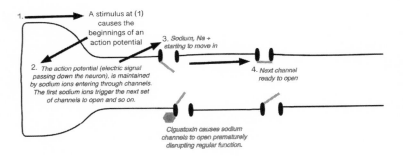

A diagram of a specialized nerve cell, a neuron. Under normal conditions nerve signals pass left to right along the cell. Ciguatoxin opens voltage gated sodium channels before the body has sent a signal, causing problems in the system. Long term disruption damages the neurons.

Things in the ocean that can kill you

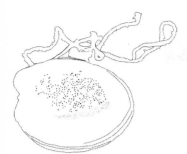

Microscopic dinoflagellates, like this one, in the genus Gambierdiscus, can build up ciguatera toxin. In turn, even more toxin builds up in things that eat Gambierdiscus because they eat millions of them. Speaking of size, these micro algae are larger than many at about 80 microns across (0.08 mm or about the width of the edge of a piece of paper).

Unless you like to surf or snorkel in shark infested waters, the most likely way that you'll encounter deadly things in the ocean is by eating something. Be alert to what you eat from the ocean.

Things in the ocean that can kill you

Things in the ocean that can kill you

Pufferfishes

There is a Japanese fish dish, called Fugu, that can kill you. The dish has other names in other countries but it is just as potentially deadly. What makes it dangerous is the concentration of a deadly toxin, tetrodotoxin. Japanese chefs that prepare the dish are specially trained to remove the most toxic parts of the fish.

Tetrodotoxin was first discovered in pufferfishes, and is named for this family of fishes, the Tetraodontidae. It has now been shown to be prevalent throughout many ocean organisms and its actually produced by bacteria that are housed in the fish.

Many species within this family are poisonous but the most dangerous to eat is the Tiger Pufferfish, *Takifugu rubripes*. Other pufferfish species, the Grass Puffer, *Takifugu niphobles*, Panther Puffer, *Takifugu paradalis*, and the Black-backed Puffer, *Takifug stictonotus*, are also deadly and among the most eaten species. Related fishes include the Porcupine fish and Balloonfish, which both have large spines to protect themselves from predators. This suggests that they do not carry the toxin, anyone want to try eating one to find out. (I'm joking, DON'T DO THIS!).

When threatened by a predator these fishes expand, by sucking in water, to make themselves harder to eat. You must be a desperate predator to chance a nasty toxin and/or nasty spines, and still try to eat them. When they puff out they can't swim very well and are easier to catch. Be careful if you do this. Though slow, the spins are very sharp. I have the unfortunate experience of getting stuck by these spines. I picked up a slow swimming, puffed up, one with a gloved hand. I then dropped it but caught it in my other, non-gloved, hand. Not my most brilliant moment.

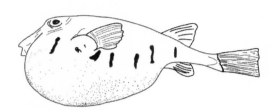

Here is a puffer, semi-puffed. When they are fully puffed they look ridiculous.

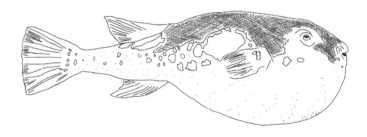

Takifugu rubripes, the Tiger Pufferfish, doesn't look dangerous but it is one of the deadliest organisms in the ocean.

Things in the ocean that can kill you

Red Tide Algae

Sometimes small things are deadly. You might not think that small, smaller than small, tiny, organisms, like the ones we will be seeing here, can be deadly but they come with friends, millions of friends. *Karenia brevis* is a single celled algae classified as a dinoflagellate; because they have a whirling (dino) whip like structure (a flagella) that helps them move through water. These microscopic algae, when sparked to reproduce quickly—called an algal bloom—can flood the water with toxins. Each of these tiny beasts holds an equally tiny amount of toxin that disrupts the nerve signal inside animals. Fishes, and larger animals, including people, exposed to too much of this toxin get sick and well, you know, sometimes die. The algal blooms are often called red-tides because some of the species that cause them have red or orange pigments and the water appears as if someone spilled red paint.

    Be wary: Water with blooms isn't always red or orange because some species of this group don't turn the water red during a bloom but instead have green pigments more like land plants.

    How small is small: *Karenia* brevis is about 30 microns wide. Let's look at the tiny measurements of the metric system again, a micron is 1 millionth of a meter or more than 25,000 microns make up an inch. Tiny sizes are hard for me to imagine, maybe this will help; a piece of regular paper is about 100 microns thick, therefore about 3 individual *Karenia brevis* cells would fit end-to-end across the edge of the paper.

Things in the ocean that can kill you

Animals, like clams, ingest millions during an algal bloom because they feed by filtering the water of these tiny cells. If you eat the clam that was harvested when a Red-tide bloom was occurring, you get sick, or worse. The sickness is called paralytic shellfish poisoning.

Human activities, particularly fertilizers used on land that get washed into the ocean, increase the occurrence of Red-tide blooms. In the U.S. this connection is underscored by the now near constant blooms that occur on the Florida Gulf Coast. This is directly connected to runoff from agriculture that first runs into Lake Okeechobee (which also has a major problem with toxic algae) and then runs into the ocean.

Be very careful, it is possible to get sick drinking, or swimming in, water with an algal bloom because you consume so many micro-algae in a small amount of water. In addition, the toxins can be released into the air when the cells are broken and so breathing near infested waters can be dangerous. Be even more careful if you have a dog. Dogs are a common victim of toxic algal blooms, particularly in freshwater, because they readily drink the water. In the U.S toxic algae are likely to be the cause of hundreds or thousands of dog deaths each year—unfortunately finding actual data is challenging as most deaths appear not to be reported.

Things in the ocean that can kill you

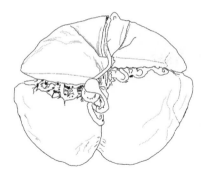

Karenia brevis is just one species of microscopic organism that cause harmful algal blooms. Its flagella are wrapped partially inside grooves. The flagella produce a spinning motion that thrusts the single cell through the water. Three of these would fit side-to-side across the edge of a piece of paper.

Alexandrium sp. also cause harmful algal blooms. These vary in size but they are mostly a little larger than Karenia brevis. This species has flagella that are wrapped in grooves and also extent outward. The test (what the shell of these single celled creatures is called) forms a marvelous pattern.

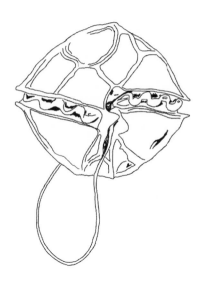

Things in the ocean that can kill you

Things in the ocean that can kill you

Honorable Mentions

There are many more dangers of the oceans. The most deadly continue to be, and most likely this danger will be increasing, are things we eat from the ocean. We fish up deeper fish than we ever have before to satisfy our plates. But these fish come with dangers. The deeper the fish the more alien its physiology. Oil fish, for example, AKA Escolar (*Ruvettus pretiosus*), contain lots of waxes in their body, many of these waxes are in the form of esters that are indigestible. This cause cause oily diarrhea in some people. The best part is that the diarrhea is orange. Orange oily leakage—yet for a time this fish was served as a sought after meal in U.S. restaurants.

In deep sea species proteins in cells can become unstable due to the pressure. A chemical called Trimethylamine oxide (TMAO) appears to stabilize proteins. The deeper the fish live the more concentrated (TMAO) becomes in their blood and tissue. TMAO is, you guessed it, toxic. So the deeper the fish, the more toxic. So as we deplete the oceans of readable available fish and search deeper and deeper to fill the desire for fish in the restaurant, well you can see where I'm going with this. We need to stop sweeping the ocean with massive nets and catching everything. This causes the plummeting of fish populations as they are unable to sustain themselves.

In Iceland they treat the meat of the Greenland Shark, *Somniosus microcephalus*, which is a deep water species and thus has high levels of TMAO in it, to make it edible. In reality the treatment hardly makes it edible. Its makes many eaters gag. The process requires fermentation and a month's long drying process. If that isn't enough to prevent you from eating this fish, consider that this long-lived and very slow growing animal would quickly become extinct it was ever a sought after menu item in restaurants throughout the world.

Plastics: The level of danger due to plastics is of great concern. We know that marine animals, everything from sea turtles, to birds, to whales, have died from ingesting large amounts of the plastic pollution we have dumped into the oceans. The concern for our health is primarily micro-plastics. Manufactured tiny beads, or minute particles broken off from larger plastics. They are in our seafood, mixed with the sand on marvelous looking beaches, and now they appear to be showing up in other things we eat and drink (including the water). How much danger these micro-plastics will inflict is unknown. An experiment where humans are the guinea pigs.

Things in the ocean that can kill you

Sources and Further Reading

Many drawings contained herein were drawn based on images available from various sources including, of course, the internet.

Barnes JH. 1964. Cause and effects in Irunkandji stingings. Medical Journal of Australia 13: 897-904.

Bouchet P, and Rocroi J-P (Ed.). Frýda J., Hausdorf B, Ponder W, Valdes A. & Warén A. 2005. Classification and nomenclator of gastropod families. Malacologia: International Journal of Malacology, 47(1-2). ConchBooks: Hackenheim, Germany.

Fields RD. 2007. The Shark's electric sense. Scientific America. http://faculty.bennington.edu/~sherman/the%20ocean%20project/shark's%20electric%20sense.pdf

Flannery A. An Ocean for Sharks. Shark-heads need to check out this blog on sharks. https://oceanforsharks.wordpress.com/author/dundunt/

Greenwood PG. 2009. Acquisition and use of nematocysts by cnidarian predators. *Toxicon* 54: 1065-1070.

Hallegraeff GM. 1993 A review of harmful algal blooms and their apparent global increase. Phycologia 32: 79 - 99.

Holford M, Puillandre N, Modica MV, Watkins M, Collin R, Bermingham E, Olivera BM. 2009. Correlating Molecular Phylogeny with Venom Apparatus Occurrence in Panamic Auger Snails (Terebridae). PLoS ONE 4(11): e7667.

James, DB. 2010. Marine poisonous echinoderms. Fishing Chimes 30: 39-41.

Jouiaei M, Yanagihara AA, Madio B, Nevalainen TJ, Alewood PF, and Fry BG. 2015. Ancient venom systems: A review of cnidaria toxins. *Toxins* 7: 2251-2271.

Lillywhite HB, Babonis LS, Sheehy CM III, Tu M-C. 2008. Sea snakes (Laticauda spp.) require fresh drinking water: Implication for the distribution and persistence of populations. Physiological and Biochemical Zoology 81: 785–796.

Lillywhite HB, Sheehy CM III, Heatwole H, Brischoux F, and Steadman DW. 2018. Why are there no sea snakes in the Atlantic? BioScience 68: 15–24.

Lillywhite HB, Sheehy CM, III, Sandfoss MR, Crowe-Riddell J, Grech A. 2019. Drinking by sea snakes from oceanic freshwater lenses at first rainfall ending seasonal drought. PLoS ONE 14(2): e0212099.

Man-Tat L, Manion J, Littleboy JB, Oyston L, Khuong TM, Wang Q-P, Nguyen DT, Hesselson D, Seymour JE, Neely GG. 2019. Molecular dissection of box jellyfish venom cytotoxicity highlights an effective venom antidote. Nature Communications 10: 1-11.

Reid, HA. 1975. Epidemiology of sea-snake bites. Journal of Tropical Medicine and Hygiene 78: 106-13.

Schlesinger A, Winter E, and Loya Y. 2009. Active nematocyst isolation via nudibranchs. *Marine Biotechnology* 11: 441-444.

Marine Dynamics Blog, http://www.sharkwatchsa.com/en/blog/

Oguri M. 1964. Rectal gland of marine and fresh-water sharks: Comparative histology. Science 29: 1151-1152.

Schulz, JR. I.Jan, G. Sangha, E. Azizi. 2019. The high speed radular prey strike of a fish-hunting cone snail. Current Biology 29: R788.

Southcoot, R.V. 1963. Fatal and other stinging by sea—wasps. Bulletin of the post-graduate committee in medicine, University of Sydney 18: 72-73.

Smith J, Connell P, Evans RH, Gellene AG, Howard MDA, Jones BH, Kaveggia S, Palmer L, Schnetzer A, Seegers BN, Seubert EL, Tatters AO, and Caron DA. 2018. A decade and a half of Pseudonitzschia spp. and domoic acid along the coast of southern California. Harmful Algae.

Truc TH, Pereira P, Mulcahy R, Cullen P, Seymour J, Carrette T and Little M. 2003. Severity of Irukandji syndrome and nematocyst identification from skin scrapings. Medical Journal of Australia 178: 38-41.

Things in the ocean that can kill you

About the Author.

Bruce has a M.S. in Marine Biology and a Ph.D. in ecology. He has been teaching and doing research for more than 25 years. His teaching spans, 2-year colleges, 4-year colleges, and field teaching with all age groups. He most enjoys field teaching and has lead students on trips to Jamaica, Bermuda, and Belize. He is enthralled by all aspects of biology and has published research articles in marine science, ecology, malacology, invasive species biology, human behavior, and physiology.

Made in the USA
Monee, IL
29 September 2019